America's Quick Fix

Danny Langley

authorHOUSE®

AuthorHouse™
1663 Liberty Drive
Bloomington, IN 47403
www.authorhouse.com
Phone: 1-800-839-8640

First published by AuthorHouse 12/21/2009

ISBN: 978-1-4490-6151-7 (e)
ISBN: 978-1-4490-6150-0 (sc)

Library of Congress Control Number: 2009913352

Printed in the United States of America
Bloomington, Indiana

This book is printed on acid-free paper.

Im like most Americans wanting a new way of life here in the land of the free without the high price tags, greed and big business dictating every part of my life. I have my own apotheosis of what needs to be done before most of what I see is forced up on us anyway. This book will make you think on the what if aspect of how we could change America and the whole world.

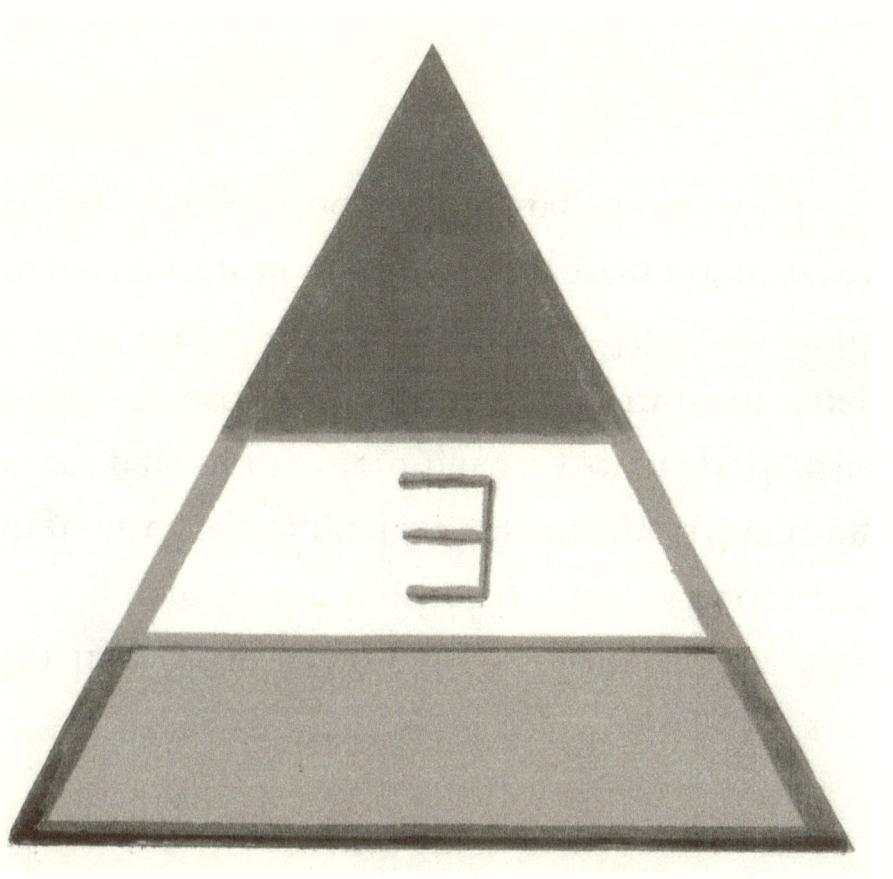

Table of Contents

Put In Place An Atmospheric Tax   1

Make Mandatory Green Laws   9

Limit Child Birth For All Races   13

English Is a Must or Get Out   17

Change Some Rules On Gun Control   21

Get Rid of Legal Tender   25

Make One Religion For America   31

Know Who God Is   37

Put In Place An Atmospheric Tax

Its an exciting world we live in and is really a convenient one. But there is so many things that need fixing, let's start with America first. And bring the rest of the world on board in the near future, we are suppose to be the leaders right.

The US citizens have paid higher and higher taxes each year after year. We all know it's all about the money. Well, when do we get a break, seems like never. The time is right for the people to make the politicians that work for us do their jobs right, make changes that count. Our government, the only one we have, keeps setting the standards and like a bunch of little duckies, we follow their directions. The earth, water and air is really all that matters, because it's the three things God made absolutely free for the whole world. We all need a new way of thinking and dealing with everything that's thrown at us. It's not that we have to change what's really inside our minds and hearts, but the way the world has become. There are two certain things in life, taxes and death. So what about the un-certainty, what if all US citizens of legal voting age demanded their own form of taxes, for the government to pay us, or charge companies literally, to use the air they trespass through, which is ours, same as the earth. Talk about wiping out the National Debt! I know I have to pay taxes on my property, just like everyone else. I always tell my neighbor, what's yours is yours and what's mine is mine, but don't cross that line, its mine to the moon. You've heard this, nothing is free, except the air you breathe, maybe we all could use that to our advantage. Let's think on the logical aspects of this, something out of our atmosphere that can actually pay down the US National Debt, for sure, if it's done right. The government keeps borrowing and spending, why not make them use their heads for change. A communication satellite costs about 3 billion dollars to build and put in outer space, that's on the cheap side for most of them.

It has to receive a signal through our atmosphere. From good old mother earth and send one back. About 40% of those billions of dollars go for building it, 40% for launching it, and 15% for insurance, with 5% going to the ground grew. Our government won't get tough on all these big companies that are making their own freeways in space, which are really limited area's to orbit in for the best return signal. We have to force the issue of making it safer in space, with a tax on the signal that goes through the air we all breathe and actually the whole world has to have to be here. Sure there is a tax in place on the money that is generated from these big companies, it's not enough.

How could you tax a signal? Simple, it is the digital age. These companies made us switch from an analog signal to a digital one, which used to be free. That means there is a way to figure out how to use the band width of the signal that is being converted into the digital signal from a satellite. Bit Rate, the digital numbers produced by the sampling rate per second, 60 seconds is 783,216,000 bytes. The whole world works off numbers. Hey, they figured out a way to make money off all of us, why not return the favor? You have FCC rules in place, sure, but it's payback time. Why should we have to pay two or three times to get a signal? We get charged for faster speeds on the internet, the connection and to subscribe to some websites we like. Even all these cell phone companies are screwing us by the minute to, this needs a complete over haul of their rules There is a lot of people that want to have a Net Neutrality put in place and free cable laws. I don't want all of these signals to be completely free, just cut the cost a little and charge a little to all companies that have a satellite in outer space to help America with its problems. It would be so easy to make them pay. Crap, the air we breathe is ours, within our atmosphere, so should the signal be, since that is between earth and space, man made the rules, so why couldn't we change them. This could be the best of both worlds for everybody, we all have to have it now, since we're in the digital age. Right?

Well, think about the money generated from the atmospheric tax that no one has ever thought about, this could take care of all the bull that is bringing us down, deficit, healthcare, and the homeless. You might think that this would be a big battle with big business and these satellite companies, to give up one 30th second of a second in the numbers 0 to 9, that is a tiny fraction within the 60 seconds that it takes to convert to digital. It could be a lot of bucks for our government to apply to the frick ups they made in the past. I mean pay for view has been up in outer space for over forty years, I do think we've finally paid off the inessential cost. Besides that, if there were a problem in making this tax happen, who owns the space shuttle, isn't the American People the owners. We all could put a stop to these satellite crashes in space, producing all this space junk and threating the security of life as we know it today. If our government were to say, this isn't possible to do, as you know they will, because big business is really the ones that have the say so. We could, turn to the insurance companies, to take care of some of our debt. Insurance companies have been screwing us, since they were invented. I think that's what keeps all of us in the slump were in. Did you know we all pay by the minute for the things that are insured and all the other things we use and have to have. They get paid that 15% of that over all cost of one satellite, add that up. It's time for us to demand they give at least 5% or more of that back to our government, if there's a problem in space, who has to take care of it, us. And beside that, most of the insurance risks, will never take place, cause when the satellite is of no more use or runs out of fuel, it is pushed into a different orbit and will be no risk to anything or anyone it just burns up. Now tell me, who really has the power of change, I think the space program could put a damper on all these money hungry companies, that wouldn't comply with the demands of the America people in a real easy way, just push their satellite into the burnout orbit, see-ya. Americans have the advantage for now, if we use it before other countries get what we have. A vehicle to outer space is our advantage and we need to make sure, were the only ones to keep watch, over space with our shuttle program. We can keep space from being polluted and taken over as long as we set the standards, people get on board now. Time is short, everything is moving at a faster and faster pace. Technology is a good thing for everyone, but when it is shared to help others, it also is turned around to be used for profit and destruction of its real intent.

Make Mandatory Green Laws

We all have to understand whats really happening with mankind. Greed, is quickly bringing us down and making every aspect of life, cost more and more per minute. All fossil fuels, have to be stopped from burning and polluting our atmosphere. We know that's not possible to do completely, how ever if they could be used for a limited amount of time or cut down 75%, it would help the earth heal its self. There's solar power, wind, wave, fusion and nuclear, but no, its about the oh mighty dollar again. These other forms of energy all come with high price tags, I wonder how much a new world would cost. The point being, something has to be done. Were headed down a path of no return. Why, one world, one life, us Americans, have to make it stop. We can, make congress apply, new laws to bring the cost down, to have these other forms of energy put in place for every citizen and ease the use of fossil fuels. If we apply this way of thinking, instead of just enjoying all the things the world has to offer at the moment with the use of fossil fuels, eventually people would follow suit. Imagine how it would be without such things, we would have to go back to the caveman days. I wonder how many of us could live that way, the earth is crying out for our help. Money and conveniences are getting in the way of what we should be thinking of, Our way of life. Right now man has enough knowledge of the way all these non fossil forms of energy work and could be distributed to our society as a whole. Here, cost gets in the way again. The United States electric energy grid, is decades old and literally falling apart at its transformers. Our Interstates, Highways, roads and bridges are coming apart at their seams to. You know as smart as we are, were pretty dumb, when it comes to taking care of our own. The billions and trillions, its going to cost to replace and fix our own economy is really a catch 22. We should incorporate all non fossil burning energy into everything that is replaced or is build new. It seems like were all inter connected in our one little world we have and live in, right. Lets take, a one mile radius of each city in America and start to build its own off the grid energy supply for that section of the city, charge everyone the same flat rate for its use. Once completed, inter connect to another squared mile and so on. Any excise power from the mile radius good be passed along to the next section. This would create jobs, cheaper light bills, along with a new way of thinking for the future good of mankind. Sure would piss off the electric companies, so, its time for a positive beginning. Even the roads you would travel on in this square mile, could have some source of energy producing aspects applied to them. Solar more than likely, imbedded in dividing line reflectors. The Green Way Of Thinking is a must, were the only ones with the power of change and say so at this point. I'm one person with an Ideal, just think if we combined our thoughts of how to make things of such, happen.

Limit Child Birth For All Races

There is too many people today in America and the world its self. Actually our world is over populated to the extent that it can't even handle the strain that's on its own resources, which are being plundered. You should have to fill out a form when you get married or live with someone, that sets the number of kids per ten years, to be born with in a time frame of no more than one child every 3 point 3 years. That will be plenty to keep our work force alive and replace the old. If a kid were to be born by rape or incest, they should ultimately be put up for adoption and let the couples which are productive members of society have them for adoption. Lesbians and Gays are perfect candidates for these unwanted children, because they are sensitive and caring people. There would have to be new laws for any unplanned pregnancies, that comes about by casual sex to be a crime, for both parties involved, in which they would be taxed to the limit, till that child is of legal age, you might make some of these teens and women, who have to be pregnant all the time, actually have safe sex. I know these suggestions seem far fetched and un constitutional, but is this the way we wish to continue. Were losing most of our rights as Americans, as another child is being born every so many seconds, a drain on society, and mother earth, our life line is being condemned to a choking death. You ever watched a bunch of ants at work, they keep building and filling their mounds with the things they need to survive. They have it pretty good like us, but there comes a time when all they worked for and saved is destroyed, because of the way they build their commuities. It caved in on them, with the over population and bad planning of their future. You see, this doesn't mean they didn't have a plan, they all were inter connected to one another's under ground nests, but in order for them to stay alive they had to move. Where are we going to move to, when it all caves in on us. There must be a way around some of the laws that are in place or a way to change them. I know we are bond by what our for fathers created for us, but its not working anymore, its being used against us. America the melting pot, were being melted away that's for sure. We took this land from the Native Indians, molded and shaped it into a Great Nation and now were being taken over ourselves. Were not the majority anymore, but the minority.

English Is a Must or Get Out

There's problems with a lot of the laws in America, we almost let anyone in our beautiful and free country and help them to become US citizens. Then they try and push all their religions and ethnic backgrounds upon us. To make our little slice of earth theirs. Once there legal, they use our own laws against us for personal gain. I know culture is a good thing and should be promoted. But, English in America has to be the one universal language for the world. If we are to make it better for America, the US must be the starting point for the whole world to follow. Something has to be done about the illegals coming here and crossing our borders. They come here and the first thing you would think, they would want, would be a job. Wrong, its to get a legal citizen pregnant and try to live off our eroding economy. Color or race should not matter as long as you are one of this society. Any person from any country that comes to America must know and speak our simple English or go home. Every time I'm out and about some where, I come across different people who can't understand, or have simple dialog with me. And there US citizens, that should be against the law here in America, they should be fined. What needs to be done, is make anyone applying to live in the US take a test to see if they know our language first, if not make them come back in a couple years and retry, not live here first. But-but, Me-me speak no English, frick that. You ever noticed when you buy something it has two or three different languages on it to explain how to use it or put it together. I'm not going to learn a new language, I was born here, why change who I am. Heres how to solve the whole issue of foreigners, just be one of us. You know if we were all one in the same, Most of the problems the world is facing would go away. Americans need to be the leaders again, not followers, we have lost so much respect from other countries and even some of our own don't believe in us or honor us anymore, its time for a change now. Not a turn for the worst, but a new beginning on how we think and see the world. It must start here, too much killing and crime at our own door-steps. A lot of that has to do with non Americans and gangs, wanting that easy money. Our population needs to slow down for awhile to let economical growth catch up. Everyone says they love America and wants to live here, that's fine, but we have to stop other people from coming and populating this great place for just a little while. We have to make them believe in our quest for our own futures sake. Who's to lead if we don't, were always helping other countries with humanitarian efforts and trying to give them what they need to make it in life. Lets take care of us first, make things better here in America and then help the ones that want to follow our since of direction and way of life. That way they will have what we all have, a good and productive life. Most of the countries we help are poor, that's for sure, but were helping them in the wrong ways, instead of just dropping off a bunch of food and supplies, Bring them some birth control first, then top soil and seeds, dig a well, make it run on solar power, give them a few chickens and a pig or two and I bet they would be happier doing for their self's, then being dependent on a small portion of something that rarely reaches them in the first place. The cost for us would be a small fraction of the billions being spent every year on a program that's not working. They just need to be educated in taking care of their self's and a little help with stopping the strong from taking what they have. Isn't that why were at war. Or is there another reason, I think not, maybe oil. If we had our own pilot program in place for America to show other countries how to limit population growth and bring on economical growth at the same time, they would eventually follow suite and their life's would be more productive and futuristic, once we showed them the way.

Change Some Rules On Gun Control

Look I believe everyone must be able to protect what's theirs and the N. R. A. along with the right to bare arms. There are other ways to keep our guns and keep them out of the hands of crimials and kids. Right now there is RFID technology called Verichip, its digital and works on radio waves. You implant a chip under the skin in a persons hand or arm, it has gps capabilities and can be used for banking and other money transactions for identification. The chip could be installed in most firearms to match the chip of the owner. That has its problems, because the guns, still could be used by someone else if the chip was removed from that person. They say it's the mark of the beast technology, point is there are ways right now to stop the violence. Most people won't go for this technology because of their faith and the unreliable forms of use in humans. What about this one, a DNA gun, makes more sense to me. Something mounted on the trigger of any gun, which you put your finger in by the trigger that reads your DNA before it will release the trigger to fire, you say not possible, bull, all people that have and own guns could have this installed on their guns. That would stop any illegal use, if a gun had been stolen, it wouldn't work, it would be usless. That way only people of legal age and that have registered for the DNA source of technology to be added on their own guns and ones they wish to purchase new, would have the fire power then. No more little punks with guns that kill. I wonder how bad they would be then, were going to have to change our ways and some laws along the way. I think this would help with a lot of the race issues to, if you take away the power to be a bad ass with a gun, there not so bad anymore. Maybe if we went back to fist to fist again, everyone would kind of respect one another. There's too many gangs with guns out there, if we apply this technology and way of thinking, all those that deal in illegal firearms would be out of business. No way can you beat DNA. The same type of technology needs to be used for our armed forces and any arms sold to other countries. Every service men and women are issued a weapon, no more losing or missing weapons. Every gun could be accounted for, even our local law enforcers could be safe in their own homes, no more accidental discharging of weapons that could hurt children. If you take the guns from the robbers, its hard to rob someone right. No more easy money for the lazy ones, they might have to go to work. This has always been a very big issue among Americans and we all know why, the right to protect ourselves. Well were really out gunned right now, this would be a form of gun control, but to our advantage. The up standing legal American would be the only ones with the right to protect his or her self. These DNA guns could revolutionize the US by making our country safer and more accountable for the actions of a person. They say guns and drugs are destroying our society, well if we could make it where only the good guys had the guns and the bad ones had no way of using them, wouldn't this make our life better. On the drug issue, why not start with legalizing pot first and see how it goes. Most people believe one drug leads to another, that's not true. It's the mentality of the person using that drug, there's a dumb ass born every day and some that just become that way. Point being, you have good and bad in every aspect of life, but if you could stop the bad guys from selling drugs to kids and let the ones that need pot to help with their pain or what ever it might be. This would be a way to help states recover from their debt and also stop all the bad things that come with drug dealers, bad parts of town and prostitution, destruction of our youth and illegal guns.

Get Rid of Legal Tender

We all know why people are robbed, its for the money. Most of the thief's are out of work little punks and don't want to work, just sell and buy drugs and are looking for their quick fix. You know if there wasn't any money for them to still or receive from the selling of drugs or stolen property, they probably would have to find a job. Even prostitutes would be out of business, without cash. That makes no sense or does it, how would they get paid? A lot of us already get paid electronically, we go to banks or ATM's and get cash, if there was no cash that we could get, what would replace it? I say again, it is the digital age, RFID technology, the Verichip. A micro chip for your body, wave and pay. The majority of people don't even know of this new technology. Is money really the root of all evil, nah, you got to have it or something that can represent it in value. Well I don't believe in a mirco chip in someone's body, that is going to far, its like I said all we have in this life is air, water and the earth, along with our body and soul. There is other ways to replace the oh mighty dollar. We all have credit cards and bank cards that work on a magnetic strip, this is about to change to that new form of technology, RFID, radio frequency identification, its not full proof yet, but it is used in every day life at this point so far, shipping, transportation, animals and the tracking of all sorts of materials. They're going to make it happen where we all are going to have to accept this as part of our lives as a form of money exchange if we let them. I told you of DNA, this is the way we have to play. Everyone has their own distinct design. DNA must take front stage, I'm pretty sure there's a way to make your bank and credit cards DNA compatible. Look, I'm who I am as you are, this has to be a must situation if we are to stay free and not be controlled by our government, to the point where they know where we are and what were doing at all times. I would suggest your bank card have a place on it to put your finger that reads your own DNA which only you are able to use when purchasing items that will show your picture at the register along with a security question only you know. I mean how hard would it be for all stores and banks to convert and make a electronically DNA format. Think of the benefits of this technology, no more cash, everything is accounted for, and no one can use or steal what's yours. Sure the government would know of your spending and deposits, but hell they know most of that now. This would make our world a much better place and safer one. No more drugs and guns could be purchased with out a record and even the cheating husband or wife, would have to account for their actions if they wanted a little on the side. Take the cash away and the bad boys can't play. So you say what about the garage sales, well I'm pretty sure a type of scanner could be invented for home use to deposit more money in your bank account. Let's face it, big brother wants his share and that's fine, we all know how it really works, you have to give some to get some. Its going to happen, one way or the other, now which way do you want it. We all have the say so at the moment. But if we let things happen with out intervention, its nobody's fault but our own. I would suggest, all American citizens take their rights to vote and use them this way. Everyone of us has a cell phone or most likely. This is the way to make our Congress do what we say for a change, use the cell to vote on the issues that can make a difference. I know you would think since even when we all voted on who was to be our president the voting machines didn't work correctly, so how can a cell phone be better. Well everyone does have a social security number, which you must give to have your phone. It is the digital age, no more cheating, call a secure number like the banks have and that phone could read the vote you put in place and know it was you by the security question that was mailed to you from the United States Congress. This could be a positive form of voting and controlling big business, finally, with our own minds.

Make One Religion For America

It's suppose to be about the people and the way they think and believe, but all of that is changing fast. That's why all religions and races are fighting here in America and around the world in one form or another. This needs to be changed starting here first. Everyone has to have faith of some kind, maybe there's the problem. Too many different thoughts that clash with the one real meaning of faith, loyalty. There has to be one faith in America first, since now that we've tried inter acting with different faiths and races, why not come together has one. We could call our selves, The American Earthen Religion, start with the mortal qualities of all religions. See how each are similar and write a new Bible or Torah for the whole world to eventually follow. I mean wasn't it a bunch of Earthens that wrote both these famous books anyway. Were all mortal beings, its just that our views are different in the way we all believe. So if we had a new religion take shape to show everyone how to make it in the only world we have with loyalty to one form of faith. The fighting and racism would go away and we would be one in the same. No more looking at color or creed, it would be just you and me. You probably have 40 or 50 different religions here in the United States alone, most of them are really tax evaders, they branched out by using our laws and forming their own religion. That's part of the problem, personal gain, has always been a factor in religion to take care of their folk. The deal is were all in the same nest here in America. Here's the law factor again, I'm telling you there must and has to be a way to change a lot of these laws, with out hurting everyone's Civil Rights. I think its really up to us Americans and not Congress anymore. Everyone talks about the Gross Domestic Product and the stock market, when it's really about if your rich or poor. No one here in America should be without housing or something to eat. Lets take care of our own first, by forming a new way of thinking and believing. To show the rest of the world our American Earthen Religion works. Its not to take away your own believes and religion, you could pray or congregate in your own homes, just leave it at your door step. When I was a kid we said the Pledge of Allegiance with God included, here's the problem to many different beliefs. There is most definitely a God, were all not seeing him in the right manner. Everyone is seeing what their hearing, gloom and doom, Armageddon. You know how it is when your down on luck and seems like the whole world is against you. All you really need is a new way of thinking to pull you out of your slump, it has happened to all of us from time to time. Look people are scared of change, that's why its slow to come, beliefs that have been handed down over thousands of years have brain washed the whole world. Most of the belief's come from what was written from a Earthen who had his own views and visions of what was to come. Were fighting with people of their own beliefs' and cultures in a foreign land, if you have nothing and bombs blow your house up, wouldn't it actually make them want to follow any type of religion, just to get back at the ones who took probably the only thing they had left. Why not change the way were handling these wars, take that half million dollar bomb, I said one bomb and apply its cost to snap together cubicles made in America to be flown in and put in place 5 or so miles outside the war zone. Connect water and power, even if you had to fly the water in. Give these poor children a chance without destruction, with the cost of one bomb. Its suppose to be about the children, you give all the young a chance to try what we have, computers, games and fast food, along with the right to be safe in their own bed at night. They would follow a new form of religion and way of thinking. You think we could take a small fraction of our war machine, and give it a try, kids really are the future. Americans have to rethink about our own slice of this world, change religion here first so there is no more prejudice on who God is. And stop the race hate issues by making it a crime for any person who did not accept the red, white and blue faith of the new Earthen society. This would stop the violence and bring on new forms of laws that could help the poor and pay down America's debt with the tax on big business. This new religion also would set the standards on Green Laws and Birth Control. Everyone would have to speak English and no more illegal guns. No more paper money to worry about, just DNA the way to pay. Help the rest of the world have a chance at what will eventually become a one for all world. We all know of the cross, if the vertical and horizonal lines were took apart and the vertical one was broke in half and put back together again forming a upside down triangle, it could be the new symbol for the American Earthen Religion. The two top points would stand for Air and water, the bottom point for the earth, these three things we all have to have to live here, which was a gift from God. Because the bottom point is referring to him.

Know Who God Is

The Egyptians were wrong about their beliefs in the building of the pyramids for their Gods, the one thing they thought was sure to get them into heaven again was the aligning of the pyramids with certain stars. If they only would have known that they were walking around in heaven at the time, they wouldn't have wasted all those years of useless work building the pyramids. You see the air, water and earth are heaven, along with our atmosphere. Were all standing on Gods shoulders, he is the core. Space is hell, its where God came from and gave us life. Black and cold like death, he and his wife arose out of a black hole in empty space as two small suns, revolving around one another. They were content being the only two of their kind in this vast dark cold place. They never touched each other as they rotated around and around, shining and staring at one another for billions of years. Oh they could communicate with small solar burst from time to time, but it wasn't enough just keeping each other in plain sight and never touching. Finally they had to know what it would be like to combine the heated passions they both felt. So as they set off one solar burst after another, getting more intense and closer each time. It happened, they touched and the feeling was so exquisite, there was no turning back. They crashed into each other, exploding and pushing parts of their selfs further and further away from one another. It seemed like it was the kiss of death for both of them. But they knew what they were doing, it was time to start their family. Gods wife stayed shining in a stationary position over seeing all the circling debris which was cast from her and her lover God. He was pushed millions of miles away from his love, but was still orbiting around her, still feeling the warmth of her love. He had been shrunk to a third his normal size, after their exploding climax. But he was still alive and ever so happy, because one third of his great hot circumference had formed their son our moon. Now God and his wife were not alone anymore, of course their son wasn't bright and shining as they both were. They didn't care if he couldn't shoot any solar flares, just happy he was there. Both God and his wife kept their son warm for a few more million years, her on one side, him on the other as God and his son orbited the one they both loved. Things started changing after awhile, the other third of Gods body that had been cast in seven different orbits were turning into growing planets. God had no Ideal that it was soon to be his faith, his body was starting to cool just enough to start collecting space matter. His shining light was getting dimmer and dimmer. But he knew he would not be like the other planets, his love was too strong for his wife and son. The other pieces of his body that had been cast into the vast black, cold, space were just too far out to feel the love. They kept growing but had no chance of ever having what he had, a burning desire to stay alive in side even as his fire was dwindling . As the years past God grew into what we know as earth. His fire never completely went out, at the core of earth it still burns. Thanks to him and his wife our sun, we have our own planets, galaxy, moon and atmothere. And through Gods love of life we have evolved, we came from a little piece of him. We swam up from the ocean floor and crawled upon his shoulders, which is the earth and have become what we are today. Earthens, we are of the earth and earth is God. Our atmosphere and everything below space is really heaven. We have to have loyalty with in our world. The point is, everyone might have been wrong about the big bang theory and the only one God we all have.

The End